Safety Has No Quitting Time

Randy Powell

RockStar
PUBLISHING HOUSE

Safety Has No Quitting Time

Published by
RockStar Publishing House
28039 Smyth Drive
Suite 102
Valencia, CA 91355
www.rockstarpublishinghouse.com

Manufactured in the United States of America, or in the United Kingdom when distributed elsewhere.

Powell, Randy
Safety Has No Quitting Time

Worthy Shorts ID: RSP119

| ISBN: | 978-1-937506-53-7 | Paperback |
| | 978-1-937506-54-4 | eBook |

Contents

Preface

This book is written for people who are directly responsible for the safety of others in the workplace. Initially my target audience was business owners but I quickly realized that every business leader should seek methods to promote the highest possible level of safety consciousness in the workplace. Whether you hold an executive position, work in human resources, or lead a work crew, if you are responsible for the safety of others it is one of the greatest responsibilities of leadership.

In the United States, approximately 4,000 lives are lost each year due to workplace accidents! Serious accidents affect more than just the employee and their family. Workplace accidents take a toll on the organization by diminishing employee morale, productivity, and the bottom line. No one wants to be the person who has to inform the spouse or parent of an employee, that a workplace injury has occurred and his or her loved one has been injured or worse. We have the ability reduce the loss of life, prevent disabling injuries and ensure that fewer people suffer from workplace accidents.

It is easy to push safety aside with all of the pressures of business, but the likelihood of something serious occurring seems to rise once we take our eyes off the ball. After investigating countless accidents, I felt compelled to try to make a difference because most accidents have one resonating theme – *they could have been avoided.*

Within these pages, you will find stories that should cause you to think differently about safety, inspire you to create a safer work environment, and provide you some information and methods about some of the new conventions of safety leadership.

As a leader in your industry or profession, you have a responsibility to others to maintain a safe place for them to work. This book will provide some insight on how to work toward building safety excellence within your organization. By leading others in safety, we show we care about their well-being and that we value them as individuals. It is remarkable how this affects productivity and is a win-win for both the employer and employee.

Acknowledgments

This book is dedicated to my wife, Susan J. Powell, for her staunch support over the past nine years, always believing that I had a gift and that it was time for me to share it with others. She supported me when I was stuck and could not write a line and she was by my side with wonderful words of encouragement through the entire process.

I am also thankful for the support of so many friends and colleagues during this journey. We all need special people in our lives with whom we can share our thoughts, hopes and deepest dreams. During motivational talks, I often remind my audiences that we have all been given the tools, talents and special gifts necessary to live out our dreams. The hard work is in taking the first step and then staying the course long enough to see that it is possible to make our dreams come true.

The journey may be a long and arduous one, but the adventures and personal growth you will experience is well worth the ride. Finally, the greatest part about finishing this book is that I get to share how wonderful my wife has been and acknowledge the awesome support and love she has shared with me. My wish is that you will learn and grow quickly from what I am sharing with you about my career in safety and leadership.

Introduction

Have you ever thought about why so many people are injured at work? There are so many other places you could get hurt so why so many at work? One day while sitting in my thinking chair it dawned on me that after working as a safety professional for over twenty years, that I have heard so many reasons why employees continue to get injured on the job. What I quickly realized is that it is our mindset that must change and it is our leaders who must get it first in order to make any real difference. As John C. Maxwell, noted author and leadership expert says, "Leaders are usually those who get it first, they are not necessarily smarter than others, but, they seem to see the big picture before others do". One of the major goals of this book is to encourage a *'Mindset Shift'* about workplace safety. My hope is that it will inspire a renewed focus on safety leadership!

I've spent a great deal of my professional life working with business owners and community leaders, many of which have grown weary of the negative trends they see within our communities. Among which is the decline in strong work ethic and sound business practices. This is a national condition as reflected by the rise in our basic healthcare and worker's compensation insurance costs. Most of my work life experience during the past twenty-five years has been in safety and risk management. It is my deepest hope and desire to use this experience to help change the traditional way of thinking about safety and risk as a necessary evil, to a more enlightened approach for today's work environment.

The simple techniques and strategies found in this book on improving relationships at work is intended to spark a new way of thinking that will result in fewer accidents, improved quality and greater productivity at work! As you read on you may find some of the stories and situations referred to may cause some uneasy feelings, others will inspire you to think differently as leaders. Safety is truly a serious business and is one of our most important responsibilities. It is so important that even now I find myself reminding you that *Safety has no Quitting Time!*

Chapter 1

Sparking a Safety Leadership Revolution

John C. Crean, in a greenhouse located in southern California back in 1950, founded Fleetwood Enterprises. He and his wife made window blinds for travel trailers and decided one day that they would build their own trailer. A few years later, he began building mobile homes and created a Fortune 500 company. Eventually their backyard business grew into Fleetwood Enterprises, the nation's leading producer of manufactured housing and recreational vehicles.

Fleetwood Enterprises was my first opportunity to drive safety from the corporate level. Most production personnel regarded Fleetwood Corporate as a necessary evil they had to endure. Corporate employees carried the stigma of being disconnected from the manufacturing group. We were viewed as working from the distant ivory towers, and so we were not part of their manufacturing team.

Furthermore, the consensus was that Corporate sat on their laurels and merely slowed down the production process. Unlike the manufacturing team, the team that gets things done! My role as a Senior Safety Engineer was to inspect our facilities for compliance, conduct safety and health audits and document my findings. It was not surprising that my role often placed me at odds with production.

The corporate office and the 'Bounder Motor Home Plant' were both located in Riverside, California. It was about 4 pm and I had just completed one of my first safety audits across town at the 'Bounder' plant. Since the audit was over, I stopped by the Production Manager's office to brief him on my findings.

It was customary to brief the General Manager or Production Manager before leaving the plant. We did this as a courtesy so they would not be surprised by the inspection report they would soon receive from corporate. Whenever the plant failed an audit, we would hold a formal exit meeting with key personnel. Exit meetings provided production personnel with an opportunity to clarify our findings which cleared the way for correcting any deficiencies.

In this case, they had passed and after a short debriefing with the Production Manager, I was anxious to get back to my office to write the report and close out the inspection. The report would not be a lengthy one as the audit had gone well. When I mentioned that I was headed back to finish the report, the Production Manager shouted back at me and proclaimed, "Nobody at Corporate ever works past 4 p.m. and especially on Fridays!" Looking over my right shoulder I replied, *"Safety has no quitting time!"*

My office was a simple one, equipped with only the necessities to get the job done. The corporate office was a two-story facility and was the backdrop for some of the hardest working and most imaginative people ever known. There were no ivory towers or corporate bigwigs sitting on their laurels back at the corporate offices, just a bunch of hard working people!

Later that evening, while in my office working on the report, the telephone rang out. At first, it startled me, but once I collected myself I answered it and to my complete surprise it was that same Production Manager. I don't know if it was a test of wills or verification of truth, but he was astonished when I answered. He asked, "What are you doing at the office?" Once again, he heard me proclaim that, *"Safety has no quitting time!"*

Even assuring him the report would be finished that evening, he was still certain that I would leave for the weekend and not get it completed. The report was finished and published as promised. The Production Manager learned a lesson that night…that corporate personnel actually do work late. Working late showed him we actually cared about the problems encountered in the field, and we could provide solutions to improve their workplace conditions. I was eventually accepted as part of their team despite working at Corporate!

Years later, we both had opportunities to tease each other about that night; we often chuckled over how it was the foundation that allowed us to build a great relationship. That particular production manager was a formidable part of creating the catch phrase I have used throughout my career and have chosen as the title of this first book – *"Safety has no quitting time."*

We both learned how important it is not to rush to judgment about others and always do what you say you are going to do. Even more important was putting in the extra time to build powerful and lasting relationships.

The "Bounder" motor home plant was one of more than twenty facilities within the organization. Our newly developing corporate safety and environmental team had the patience and developed that same level of trust at every one of our facilities. It took a significant amount of time and effort to convince production personnel that corporate really did care about their challenges and that we didn't just sit around in those so-called ivory towers. Together we were able to change production's mindset.

Even though small businesses are the backbone of our economy, they lack much of the support that larger corporations are able to secure. There will always be challenges, but the new philosophy has to involve taking a closer look at every aspect of the business because it is critical to their long-term success. Anything that can ease the many burdens and lower the cost of doing business must be done. There are companies that are fanatics for safety and they are doing a great job of minimizing their risks. They use their 'Best-in-Class' safety performance records to their advantage.

Companies often generate new business because they promote safety as one of their core beliefs. Doing that allows them to market the public in ways that benefit the company. There are still many businesses that don't understand the power of a 'Best-in-Class' Safety System or understand how to build it into their culture.

When corporate executives at Fleetwood finally understood the broader impact and the possibilities that existed for those who reach the higher echelons of safety performance, they insisted on it becoming part of our strategic plan. Therefore, it became part of Fleetwood's new culture shift with a change in mindset that began a new era of safety leadership in an ever-changing and dynamic corporate environment.

If you're responsible for the safety and health of other people, you must decide to stand up for safety. It is a philosophical matter as much as a moral obligation. Anyone who refers to themselves as a safety professional understands this because they are specialists in their field; someone must stand up and ask our leaders to lead differently today.

Things won't change without a dramatic shift in leadership thought. The truth is that it's not always easy to stay the course, in fact most of the time safety and human resource professionals feel like salmon swimming upstream against the current. Often times managers struggle and find, they are caught between deciding to do what's right versus what's riskier to get the immediate short-term results they want or need to show progress.

Every manager should take the safety of their employees personally; the pain and suffering that workplace injuries cause extends well beyond the workplace and the effects are felt and shared within our communities. Starting the conversation about changing how workplace safety is viewed is just the beginning. By sharing our experiences, philosophy, and the leadership lessons that we have learned, we could spark a revolution for safety excellence and get it into the mainstream of corporate America.

It's all about caring for people and the quality of their work lives as well as their families, friends and loved ones. When we show great care for one another, we all benefit. A change in the mindset of those who want to manage and lead others in the future may give us an opportunity to look back and point to a life we've saved. That is worth my time, having something tangible that we can leave behind, isn't it worth your time as well?

If you have children, the next question will be much easier to understand. However, if you're not a parent or guardian, there are many situations in life that will allow you to thoughtfully consider and answer this within your own mind. How do you prepare a child to work in a dangerous world?

After many years of working as a safety professional, in various industries, one simple question remained on my mind about my own children. How should we prepare them to work in a very dangerous world? What types of conversations should we have with them that might be the difference-maker and keep them as safe as possible?

One day they would be exposed to tremendous hazards and forces beyond their wildest imaginations. The more exciting or thrilling their occupational choices were, the greater the chances they'd experience some of the pains we know which result from being part of this great nation's workforce!

The greater the exposure, the more exciting and rewarding their work experiences will be to their personal growth. Growth is extremely important and experience is the best teacher! So the more they are able to experience, the more meaningful and engaged their work will become and the more valuable they become to others.

Perhaps the best answer is simply this; prepare a child to work in a dangerous world the same way that we would prepare ourselves. That is to build safety into a lifestyle, whatever lifestyle you chose. A person can always consider that ___*Safety Has No Quitting Time*___ and apply it to their lives in a manner that works for them.

My desire for protecting people, coupled with my life experiences will create a shortcut for leaders who want to lead successfully in all areas of life. It is my desire to help people live completely fulfilling lives, but to do this requires a considerable amount of personal effort to act and work responsibly. People must be willing to weave safety into their lives both at work and at home.

Today's work environment is looking for the next generation of leaders as well as those who are in current leadership roles today. If you're reading this book, someone recognized that at the very least, you care about people and perhaps you have what it takes to become a better leader and influencer.

This book is about increasing our awareness about the importance of workplace safety, its effect on our lives and the improvements that we can make to have a more productive and fun-filled work life! This will make work less dangerous for everyone. We've all experienced it with our own eyes; people rushing to do work, engaging in commerce and too often the rush becomes blurred. When tragedy hits, and it will hit, life stops and people suffer and heal. Sometimes not so well and their lives are forever changed.

Then the wheel of commerce begins to turn ever so fast once again and we soon forget about that last incident that affected someone. Most of us don't want to forget, but the trouble is we do...and as they say, life goes on, or does it?

As leaders, we must take the opportunity to continue to change our philosophy about work...and safety...and the risks we take every day! We ought to remind ourselves, our family, friends and the people we live around and interact with daily, about the very powerful truism that, ___Safety Has No Quitting Time!___ It must become part of our life's philosophy, so that we continue to live the lives we were born to live and reach our fullest potential!

Imagine living in a world without any regard for safety. How would that impact productivity, innovation and creativity? Then one day as if out of nowhere we receive that fateful phone call, the one call every parent or person that cares for another fears! A tragic accident at work has just occurred and your loved one or dear friend was involved.

Sparking a leadership revolution in safety will require new thinking, the will to act and the courage to stand up against the many forces that people find themselves often losing out against. It will take more than a single voice to speak and generate change in business, government, schools and within our communities.

The experience of working for some extremely large corporations brought with it countless accident investigations that could have been avoided. Over the years, we've all seen improvements in technology, which has had a positive effect on safety. However, the greatest impact we can have on safety won't be designed or developed.

It must be inspired by leaders who understand that it's too often a lack of great leadership that drives us to such tragic endings at work and at home. We must lead correctly without regard for titles. Corporate America has the power to move at lightning speed to ensure that our future leaders; lead with an understanding that every life counts!

My message is somewhat of an appeal to raise a generation of new leaders who understand and respect life so much that they will desire to inspire and drive the changes we need to improve our work lives. It's a dangerous world that we live in, as we all know; our workplaces and the process of what we do should be designed and established with an understanding that people are our greatest asset. We owe employees a leadership mindset that shows they are vitally important to their organizations.

Nothing great is ever accomplished alone and more can be done with the fantastic teams that we are able to build! Our greatest challenge will be overcoming our own fear of change so that we can make the right decision to protect others.

Chapter 2

The 5 Greatest Safety Challenges

In some organizations, it is very difficult to figure out the greatest need at any given time and it was no different for our team at Northrop Aerospace. The greatest challenge that we faced was to understand why our handpicked team was chosen to manage the Occupational & Explosive Safety Program for a classified project back in the 1990's.

From the first moment, our department stood out from others and the reason was leadership. Our Department Manager was a tall slender man who was a very serious minded individual with one vision, to be the greatest program support team within the Advanced Programs Division. To do this he had to assemble a team of specialists and connect with each of us in a manner that would excite and drive us to excel every day! With no exception, the highest regard for security, professionalism, ethics and performance were expected.

He built personal but professional relationships with each team member and when we came together, which was also very rare, we jelled as one unit! One of the earliest lessons he taught us was how important our roles were to the program and the necessity of diversity, which would be one of the keys to our success. His vision for our department was not selfish because he understood our position and knew how important our role was to the program.

He knew without a doubt that if we raised our level of performance, the entire organization would step up their game and our program might stand a chance and outlast the Congressional budget cuts that were looming over the nation. The vision our manager painted for us was greater than our department, we understood our significance, he maintained a relationship of trust and he was a leader who allowed us to do develop a system that worked for our team.

Although the program was eventually shut down, we achieved our objectives before they closed the program. We were successful because the product we were developing was finally tested and proven to work! That made all the hard work, the hectic schedules and the long hours on the road to get to work worth it.

Every department within the Aircraft Division recognized us, treated us special, respected our work and had the highest regard for our manager. Whenever our department was recognized for an outstanding achievement by either our Vice President or the Program Manager, our manager never took credit himself and always shined the light on his team!

First Greatest Challenge – Establishing a 'Grand Vision' that people can buy into in today's marketplace can be a daunting task, especially for smaller companies or organizations. If you deliver a clear and consistent message, one that people can relate something of their values to, they begin to understand and accept their role in supporting your vision. Great leaders have a keen sense of painting the big picture and communicating so that it becomes real and an attainable goal.

At Northrop, our team was able to reach high performance levels in a unique environment with minimal daily contact because we understood the mission, our roles and we developed a strategic communication system that worked. Communicating the vision, setting up a communication process that works and captures something that everyone values can stir the imagination of a great future and inspire people to perform at their best!

Whether it's a start-up company or a well-established organization, it is extremely important to have a great employee orientation process that communicates the vision and mission of the company. Take the time to make it a normal part of the company's communications process when employees are first hired and throughout their employment. The importance of employee safety, health, and environmental matters cannot be over emphasized during the earliest part of your employee indoctrination period.

Organizations that speak lightly about these topics and bury the information inside their employee handbooks in today's work environment will not succeed in building a strong safety culture. Spending an ample amount of time with new employees will give you a better chance of influencing the type of behaviors that you would like to see woven into the fabric of your organization.

By the way, spending time with new employees gives you an opportunity to explain their roles and show them how they fit into the organization. The more that you can include employees in the process as you build safety into your company culture, the greater chance you have in achieving 'Best-in-Class' safety performance.

Remember that whether your company is a small start-up or a larger organization, employees must understand the business, what future success looks like and exactly how they fit into the organization in order to help them buy into the vision. Doing this will reduce injuries, lower your costs and boost employee morale.

To build a great vision, leaders must also find the best possible method to communicate the goodness found within their products and services. Take what Steve Jobs did with the I Phone and picture how your organization is changing the world. It may be at a different level, but the concept is the same. It doesn't have to be unique to Apple! Learn from those who've gone before you and look at what they've done to be successful in leaving their own legacy.

Second Greatest Challenge – Another one of your greatest challenges will always be your ability to work with other people. Humans are special creatures who are dynamic and have ever-changing personalities that are charged with emotions, stories and their own unique histories that have shaped who they are today.

With that in mind, perhaps our efforts would be better served looking at what we can do as leaders to enhance our understanding of what it would take to actually achieve our one great goal! As leaders, we are technically responsible for the safety and success of an organization; shouldn't we be the first to acknowledge where our success is actually rooted and how best to arrive at its doorstep?

It has been my experience that successful companies realize it isn't found solely in the product they produce or the service they provide. It is most often found in the character of the people they employ and the ongoing training and education process they provide, that have the greatest impact on their organizations.

Diversity is a major key to their success as well and it supports a growing and thriving business because of the quality, resourcefulness and multicultural advantages that it brings into their group dynamics. In a global economy, diversity is a crucial part of our ability to expand our business reach. It makes it easier to learn what is significant or important to a much larger consumer base.

Digging deep into the motives and management styles of our leaders will reveal any critical gaps in our management system that may have a negative effect on safety. Remember that sometimes the people we place in charge and have the greatest opportunity to make positive changes don't!

We have found through the years that even managers are often blinded by ego or position. In essence, they've become too close to the process and exhibit an emotional tie to the problems, which can impede the organization's progress.

The human body is remarkable and just as it has systems that sound alarms when things are causing it distress and pain, so does your work environment. One problem is that we tend to ignore the signs of stress, pain or environmental warnings until people break down and require medical attention. By the time we acknowledge our opportunity to make a workplace adjustment; it's often too late!

Here is a prime example of waiting too long to act; when employees are modifying the tools, they've been given or we see that our injury reporting processes are failing, they begin to wonder if we care about them at all. Left unchecked too long, we lose their confidence, which could breed discontent. These underlying issues go unrecognized and often ignored by supervisors. Yet supervisors are being paid to protect employees from such hazards and are failing to recognize their roles and take appropriate actions. Therefore, they are part of an already weakening safety system.

Whenever a management team loses focus, lost time injuries and medical costs begin to skyrocket. Employees begin talking negatively to one another about their aches, pains, and discomforts. Suddenly, you recognize that your employees are not recovering as quickly from their injury claims and your medical or indemnity costs begin to rise.

Eventually you see a rise in litigated claims and your organization begins to feel the pinch! Ultimately, these types of issues may cause lost productivity, poor quality and service. At some point your customers and clients begin to notice and in the end your vision for the company is hindered, your focus is sidetracked with daily bouts resulting from a poor safety program, which is now affecting the overall health and wealth of your company.

This is a huge challenge that can be avoided by assembling the right team members, making sure they know and understand their roles and then responding promptly when problems are uncovered.

Third Greatest Challenge – Building and maintaining the right relationships is something that good leaders are skilled at and they should exhibit these skills all the time. This is a learned and practiced set of skills that anyone can develop in their leadership journey. Leaders are responsible for other people, especially in the workplace; they need something from their employees and employees need something from their leaders. It's a two way street and the overriding need is trust…trust takes time to develop and it can be easily destroyed.

Good leaders, such as the one who led our team at Northrop, are always concerned about the well-being of their organization. They rarely take credit for themselves! They feel their way through an organization and often hire talented personnel. Like anyone, they have specific duties and responsibilities and even they can be overwhelmed by the enormous pressures of business. All too often even with the best intentions, executives leave their organization's relationship building to un-skilled managers. Therein lies part of the problem. Organizations should establish an effective method that allows business owners and executives to maintain a hands-on approach to building and maintaining the right type of relationships with their employees. This is an endeavor that cannot be left up to the unskilled because it is too important and over time, they'll mess it up! You must have the right checks and balances in place to establish that people are being treated in a fair and just manner.

Leaders know their own strengths and weaknesses and great leaders use this knowledge to advance their cause; in other words, they hire good people who can fill specific gaps in their organizations. Leaders must be able to admit their weaknesses and work to close those gaps by hiring people unlike them in order to strengthen the team. Adding the talent and skills we are lacking shows our leadership maturity and that builds a stronger team. We don't have to possess all of the answers, but what we must do is build great teams!

The ability to step outside of oneself and see talent, hire talent and begin building solid relationships with people is critical to the success of any organization. When your team sees that you hire well it will build trust, simply because we have all been a part of organizations where too often key positions were filled with people who looked exactly like our leaders. They hired them because they liked them, only to find out later that our team's problem was never solved.

It's been my experience over the past 25 years of working in safety and risk management, that management teams also lack real empathy for their employees. Managers can easily be caught up in the daily grind and forget their employees are people too! The best companies exercise their ability to stay close at what I call the street level.

Staying in touch with employees at the street level is much easier in smaller companies. Mid-sized and larger companies often struggle to stay in touch with employees because of their size and organizational structure. Business owners and executives often leave others in charge and back away once their organizations are large or successful enough to operate on their own. They begin losing the connection they originally built with their employees and this often causes a communication gap to form between various groups within the company. Communication is the first thing to suffer and the message from the top becomes watered down by the time it reaches your employees. This has been the cause of many problems within an organization.

There are other reasons for this but the most important consideration is whether you are able to maintain a connection with your employees. Staying connected to employees will keep them from losing sight of your vision. If you lose this connection, it may ultimately increase your risk and the likelihood of you experiencing a weaker safety posture.

A weakening safety posture will place your company in jeopardy of injury claims, higher worker's compensation costs, fraud, loss of productivity, lower morale and much more! By now, you should see the big picture that is being painted. Do you see how this all links together so very quickly?

Fourth Greatest Challenge – Installing a systematized approach to safety, health, environment and the public good is another great challenge in business. People are somewhat concerned about whether companies are just in it for themselves or whether they care about the communities, in which they operate.

Integrating a culture of safety into the fabric of an organization isn't new, but it's rare to find many smaller or mid-sized companies who have mastered this art. It takes time, it's an investment in people and it is hard to let go of old traditions. It also takes some advanced leadership skills that must be learned throughout the entire organization.

One of the toughest things to do is to ensure that a system is in place so that you're not wasting time and resources on problems that were previously solved. If you put a process in place that captures how previous problems were identified, resolved and then document the corrective actions that were taken, you would have a system and a data source to fall back on when it is needed.

As an example: the accident investigation process should be designed to identify the 'root cause' so that prevention methods can be identified and corrected. It's been my experience that we spend much of our time in the mud and we fail to work the accident prevention side of safety. One of our ultimate goals should be to eliminate problems once and for all whenever possible.

When you systematize you shorten the disruption period when an incident occurs and that allows the company to continue producing your products and services. For many years safety was a hit and miss proposition and unfortunately that's still the case in many organizations. We are experiencing a major shift among those leaders who wish to rid themselves of the painstaking choices that have not worked for them in the past.

Good companies are beginning to understand the difficulty and the challenges they face when safety is not woven into the fabric of their organization. The desire to move beyond good to great is increasing and that will be seen through our actions.

A systematic approach to safety, health and the environment puts the necessary checks and balances in place that will minimize the waste, fraud and the abuse that is caused by poorly constructed government programs such as the workers' compensation system. We all know that some employees take advantage of the worker's compensation system. Unfortunately, for many, by the time the costs have escalated high enough to grab their attention, they are already experiencing the legal and medical madness that surrounds worker's compensation. Developing a safety system that works whether or not you are present should be a high priority for any company seeking to have a 'Best in Class' in safety program.

Fifth Greatest Challenge – Government regulations are a major challenge because agencies speak a language that is very difficult for people to understand. Individuals who are unfamiliar with regulatory standards find it hard to interpret the requirements and apply them to their operations.

The rules governing the workers compensation system are known for causing ill feelings between employers and employees. Smaller companies know a lot about their employees and that can sometimes cause them to let their personal feelings get in the way of making proper decisions. Some employees have learned how to abuse the system and this has caused disdain toward regulatory agencies; most of which will either directly or indirectly affect the outcome of citations or injury claims.

When business executives, owners, human resource personnel or managers become openly aggravated with employees during the worker's compensation process, it almost always slows down the process of closing out claims. Showing any signs of dislike toward an employee for getting hurt or for filing an injury claim will likely cause employees to distrust you and seek legal counsel which further complicates the situation.

The best way to manage employees' claims is to ensure that you have correct policies and procedures and a well-balanced safety process in place. You also will need well-trained managers, staff, a system for providing timely response and the ability to give personal attention to an injured employee. Even if you suspect foul play, it's better to respond appropriately because the cost of creating more conflict in the workplace will escalate and spread to other areas within the organization.

The worker's compensation process is tedious and time-consuming, but your insurance carrier should carry the majority of the burden once a claim has been filed. They have the advantage of not being too close to the claimant and they are your experts! You have enough to do with completing the incident investigation and identifying the causes, recommendations and corrective actions necessary to prevent a similar case.

Maintaining the right attitude and mindset is an important part of the worker's compensation claims process. Over the past several years, employers have complained about a flawed system, especially within California. These types of conversations take up precious time, energy and resources. The greatest danger most companies face with regard to the worker's compensation and insurance industry is not having a sound philosophy and safety process in place that works.

Understanding what's in your control and how to manage it, is more fruitful than fighting it and complaining about the system we find ourselves forced to operate within these days. Regulations are here to stay and the worker's compensation system is flawed in many ways; the legal system isn't designed to eliminate the waste, fraud and abuse, it's simply there as means to allow you to participate in the legal process. The best advice is to get involved at the legislative level in order to seek changes to the current system.

So there you have it, the five greatest challenges facing companies today! Looking through the eyes of someone who has been on the front lines coast to coast, working in hazardous and unstable environments, inside test labs, and from manufacturing to sub-assembly plants, my findings and conclusions are the same. These are the five greatest safety challenges facing companies today! Learn to solve the puzzle and master these challenges and your company will be well on their way toward reaching excellence in safety!

What you can do as a leader has been placed here at your doorstep, take advantage of it and enjoy the ride! None of these challenges is so daunting they can't be managed. The truth is that it takes sound leadership and a willingness to understand and care about people. As leaders, we must rise above the petty things that employees may do when they feel they can't change their own circumstances.

Leaders should recognize that an individual needs far more than just a paycheck; we are much more complex than a simple payday! We're talking about your greatest asset, which is and will always be your people. To be wildly successful you have to genuinely like people, heck, if you don't like people you won't succeed!

So have the courage to invest heavily in every employee, if they are still on your team, you must see value in them so as someone once said; they only subscribe to the **'B.O.O.G.I.E'** philosophy when it comes to employees. They preach that employees must **'Be Outstanding or Get Involved Elsewhere'...B.O.O.G.I.E!** We're talking about the type of attitude that it takes to build a *'Best-in-Class' Organization!*

Chapter 3

Eleven Modern Day Risk Factors in Todays' Work Environment

Here in this section are the twelve most prevalent reasons why companies fail regarding safety, health and the environment. These eleven risk factors will only apply if your organization doesn't have them in place today! If you have addressed some but not all of these issues, you still have work to do because each of these risk factors could cause lost productivity, wasted time or precious resources and that will ultimately reduce your profitability.

First Risk Factor - <u>Poorly written or inadequate safety policies and procedures</u>

It is easy to overlook policies and procedures, especially when so many companies are often formed by entrepreneurs who may not have the time or skills for such administrative tasks. The greatest concern is poorly written procedures that may not be updated to meet your current challenges or in many cases, they are not being followed at all.

Policies and procedures are an essential part of a successful safety program because they provide specific direction to help you achieve your goals. They also make it easier to develop more advanced programs and if you include employees on your policy development team; you gain greater employee buy-in since they see their fellow employees taking part in the policy making process.

> **Warning** – It is a mistake to purchase off-the shelf safety programs and binders full of standard programs and then call that your safety program! However, if you believe this will save you time and energy, make sure that you purchase this type of product only if it can be modified electronically and you are given the rights to the product's templates. Regulations and requirements change and you want to be sure that you are able to purchase any updated safety materials and programs for a minimal cost.

Second Risk Factor - <u>Inability to follow Safety Policies & Procedures</u>

Sometimes people find it difficult to follow outdated policies and procedures when they know there's a better way to get the job done. Getting the team to follow the organization's policies and procedures is a twofold process, but most of all it's a leadership issue. One of the easiest ways to accomplish this is through an <u>imbedding process</u>. However, you must first make sure that your policies are easy to follow!

Imbedding is the process of incorporating all of your written safety policies and procedures into your standard operating procedures so they work together as a single source document. When this is done well it becomes easier for employees to accomplish their work in an efficient manner. My clients use 'Mastermind Groups' because they are a great way to tap your internal resources and use employees who are skilled at incorporating the team's ideas into written form. These teams can easily address various ways to incorporate new standards or policy changes that affect their fellow co-workers.

Using employees who you know are people of influence and skilled at systems thinking is an advantage because it takes less energy to get the job done correctly. You will also need members to understand the benefits of what you are proposing to change to improve safety. Putting together a great mastermind team for major projects is well worth the effort!

Companies have used focus groups, committees and now masterminds it seems forever. It supports a collaborative process that was originally used in the 1990's with great success! However, today's masterminds are even more powerful and bring innovative and thought provoking possibilities to the problem-solving table. These type of teams are often involved in much of the 'Imagineering' that's required in today's fast-paced industrial and economic environment.

Third Risk Factor - No Safety Budget

Budgets are always a tough conversation to have and smaller companies almost never consider safety when planning their annual budgets. This is problematic because it can send the message to supervisors and managers that safety is not really a high priority. When 'safety' is not a line item in the budget it provides an easy out for managers wishing to refrain from planning or spending the necessary funds that will support establishing a stronger safety culture.

You might be saying to yourselves, well that doesn't make sense and that may be true, but, consider the new supervisor or manager who is developing their skills and has been given an opportunity to lead people. They are not always equipped to understand how this affects the organization at this stage of their careers. It is called tunnel vision at best, worse yet, it is ignorance. They don't know what they don't know!

Organizations who understand that budgeting for safety means that you have a well thought out plan for including safety into the overall process is setting the company up for long-term success! Your budget should include setting aside training time for management, employees and for accomplishing certification programs that could help you meet any annual regulatory requirements.

Fourth Risk Factor - No Safety Process in Place

A safety process doesn't need to be complicated, however, it is easy to mistake or confuse a safety program with having a safety process. Not understanding what constitutes a safety process, which is not the same thing as a safety program, is a huge reason why OSHA audits often do not always go so well. A safety process is the roadmap or internal guidelines that an organization uses to ensure that safety becomes part of the culture. Once this occurs, the organization starts to accomplish what is required without any extra effort!

Therefore, a 'Safety Process' can be written out but is not required or mandated by any regulation. It can be described simply as the internal methods and means that organizations use to ensure that whatever is appropriate is done in a timely manner.

'Safety Programs' on the other hand are usually designed and written to comply with a Federal or State regulation. You will undoubtedly notice that I often refer to California's safety and worker's compensation rules and regulations; this is because they have been considered to be the most stringent laws in the nation and if you can be successful operating under those laws you will find it much easier to operate in other states. Written safety programs are designed to simplify complex regulatory requirements so that it is easier to meet compliance objectives. Written safety programs also demonstrate corporate compliance and they explain the documentation required or proof of their compliance. California requires that every business entity with ten or more employees have a written *Injury and Illness Prevention Plan or (IIPP)*.

The eight elements listed below are required by Cal-OSHA and this meets the requirements of a basic safety plan.

1. Responsibility
2. Compliance
3. Communication
4. Hazard Assessment
5. Accident/Exposure Investigation
6. Hazard Correction
7. Training and Instruction
9. Recordkeeping

The policies and programs that are developed to meet additional Cal/OSHA regulations are determined by your industry and specific operations. Remember that your IIPP must remain separate, as that is your safety plan. Your safety policies and procedures, often called your 'Safety Manual' or 'Safety Binder' should be maintained separate from the IIPP. The ultimate goal is to eliminate any reference to a 'Safety Manual' as you work toward imbedding these requirements into your Corporate Policies and Procedures.

Safety Supply Companies, Consultants and Vendors know this and will sell off-the-shelf programs or software packages designed to leave you with a product that fulfills the basic IIPP requirement. You should know that is just the tip of the iceberg as they say and it usually leaves companies feeling they have what is needed.

Unfortunately, for most, these programs must be customized for their operations and they must be site specific for the hazards that employees may encounter. Most small business owners don't have the time to drill down and understand the regulations that govern their operations. Therefore, it is easy to rely on vendors who sell seemingly simple solutions that sound awesome in today's hurry up culture.

Always remember the problem you are trying to solve along with its complexity should dictate whether you invest in hiring a cost conscious safety consultant or depend upon a particular vendor who might be selling an out of the box solution! Safety professionals possess specialized knowledge and they understand the regulations and that can save you time and money. This is especially true if you are continuing to experience high incident rates and cannot get control of your injury costs.

Injuries are reflected in the cost of your workers' compensation premiums. When your insurance modifier rate (mod rate) rises, the rate you pay for insurance increases. Premiums are also affected for (3) years after the loss occurs. Insurance brokers will try place you wherever they can during this time because your rates may have become too high to put you with a traditional insurance carrier. This occurs as a result of a poor safety record, the good news is that it can be prevented and your risk can be lowered if you establish a great safety culture. Controlling injuries begins with having a viable safety plan in place and a program that meets the regulatory requirements.

Next, employees must be shown how they benefit from working safely and how reducing the organization's injury costs and improving safety provides the opportunity for job security. There are no guarantees but it helps when an organization can work more efficiently, lower their costs and improves safety too! Finally, once you have a safety plan in place and it is working; it may further justify management's support for continuing to build an even stronger safety process.

The important thing to remember is whomever you seek advice from must understand the concept and differences between programs and processes. Better yet, they should have a systems thinking mindset if they want to truly prevent companies from reinventing the wheel over and over again.

Some companies fall into the trap of shortcutting the process and they often fail to correct the root cause of their problems. These are <u>opportunities for improvement</u> otherwise known as (OFI's)! Taking this perspective prompts work teams to view problems as learning opportunities as well as a developmental process rather than focusing on the negative.

Fifth Risk Factor - <u>Poor safety training or lack of training time</u>

The importance of training cannot be overstated and the amount of time a company spends on training will directly affect their overall success. For example, if you have ever visited a Nordstrom's you know right away that they spend a great deal of time on customer service training. Why do you know this? It is simple. There is a feeling that you get as soon as you enter one of their stores. You are always met with that professional and courteous greeting and their approach to problem-solving leaves the client feeling understood and heard. A feeling that is magnificent!

Their employees listen tentatively and only then try to help you solve your problem. The amount of time and quality they invest in training reflects their corporate commitment to the education and training process. Nordstrom's is known for cross pollenating their impeccable service throughout their stores and mastering this skill gives them a strategic advantage. It is representative of those organizations that seek to be the best.

Service is their trademark, it is ingrained in their culture and as a result, Nordstrom is recognized as a leader in the high-end retail-clothing sector. They are service-minded and it remains one of their greatest hallmarks of excellence. Training is an absolute necessity for higher levels of success. You can always tell when a company believes this because it is easily recognized within their establishments.

Training improves the quality of your services and it is an essential function for maintaining a high standard. _'Best-in-Class'_ companies usually devote entire departments to ensure they maintain the best possible training programs available.

Great training programs are well documented, written and published in a manner that gives employees a better understanding about how to reach safety excellence. You should continually assess the quality and training needs of your organization. Great training is a key element and in many cases, it becomes the difference-maker.

Have you ever dined in a restaurant that had a weak training program? You knew it was going to be a poor experience right away and you felt compelled to leave but did not. This type of situation always has more to do with the quality of training that employees are receiving along with the culture that has been adopted by the restaurant. It is no different for any business and that is why it is extremely important to identify the best possible candidates to become your trainers.

Some of the best trainers are those who love to teach, are patient, goal oriented and are good communicators.

Encouraging your trainers to reach a certain level of mastery is a worthy goal for any business. Nevertheless, no matter how great a training program becomes it can only be maximized with great leadership in the forefront. A good training program should also include an element of cross training. This is something used and perfected by the armed services and it is one of the best ways to narrow any training gaps that may exist. Doing this may provide you an opportunity to fill a vacancy immediately when the need arises.

Finally, if you need technical skills that you do not have perhaps the best option would be to utilize consultants until you are ready to hire specialists. When the company grows, it may become feasible to add safety, health or environmental professionals to your payroll.

Everything you do should free up your time to focus on areas that directly support the core products and services that you bring to market. When you contract for services, it is also important to remember to assign a company liaison who can oversee the program or project. This will ensure that your organization's best interests are being served and protected at all times.

Sixth Risk Factor - Low morale

Low morale affects people in different ways and it usually has an immediate, but often silent impact on an organization. Employees may not complain about being overworked but too much overtime may increase your risk of injury. Overworked employees often take short cuts and this leads injuries and discontent. If employees are forced to work overtime too often it will lower morale and cause them to feel as though they are not valued by the organization.

Employees also stand the risk of becoming uninterested in the company's long-term objectives because they may feel mistreated by the company if overtime becomes the norm. This may lead to the perception they are being treated unfairly if they are under pressure to work longer hours. Ultimately, that could affect productivity and quality.

Seventh Risk Factor - High Injury Frequency

One of the most difficult realizations for business owners to grasp is that they are directly responsible for their losses. The reasons that high injury frequencies, severity and worker's compensation costs are rising or are not under control is directly related to how well the leadership does its job! This is a difficult reality for most managers to face and they will not embrace change until they accept responsibility for their conditions.

Poor leadership can easily be turned around once management accepts responsibility for their part in the process! This is where supervisor and management training programs become vital to your organization. Training will provide supervisors and managers with an understanding of the complications associated with the worker's compensation system, managing injury claims and the legal maneuvering by litigant attorneys.

Supervisor and management safety training classes are designed to teach some unique coping skills and will show management personnel how to oversee injury claims without getting emotional. Believe it or not, following the claims process and maintaining as close a relationship as possible with injured employees is one the best ways to minimize problems with an injury claim. In most cases, employees will return to work sooner and you will get the claim closed much faster.

The solution is in the details of how well we lead when designing our communication systems, in treating and following-up on injured employees. Taking these simple steps and showing some care for employees will curb their appetite to lash out and reduces the chances of them filing recurring claims, becoming injured again or raising other problems, which will affect your injury frequency rate.

Eighth Risk Factor - Time Management

People still believe there is no time to get it all done... that requires a mindset shift and an understanding that we only need to focus on the right things, the rest will take care of itself. There is a lot to be said about time management and the truth is we always finish what we enjoy first! Therefore, we need to plan ahead of what we despise and let someone else who's better do those things we're not so great at in the first place!

Ninth Risk Factor – Either 'No' or a flawed 'Safety Philosophy'

Some managers have never considered what their philosophy is towards safety, risk or their work environment. In today's market not having a safety philosophy as a starting point will at best slow progress and cause inefficiencies and at worst it could cripple an organization.

Consider the possibilities of what having a strong safety philosophy might mean to your organization. Having the right safety philosophy and believing in its power could make it easier to implement the changes necessary to improve your business. It might also give you the strength and courage that you need to support your decisions with regard to employee safety and health matters.

Tenth Risk Factor – No Benchmarking Program

Less than 5% of the business owners use benchmarking and even fewer take the time to go out and visit other companies within their own industries. Whether it is their ego or stubbornness is unclear but it cannot be they do not know where their competition is located.

Small companies seem to focus more on short-term objectives and have little time or feel they do not need to benchmark their business. Benchmarking can be a leading indicator to the health of the industry and the organization due to changing market environments.

The more complex that an organization's infrastructure and operations become, the more difficult is it to make major system changes; meaning the more resources it will take to correct problems once they are discovered. Benchmarking can help an organization stay ahead of the curve by providing critical market insight so that an organization can plan any necessary changes in a systematic manner and remain cost effective.

Eleventh Risk Factor –No Emergency Response Plan

We know that most management teams have never had to respond to a catastrophic event! Better to be prepared than not, emergency preparations and response is a critical part of minimizing and preventing injuries and business interruptions. Your initial response should include the necessary support equipment and training that is required to manage an emergency response activity. Things happen very quickly when a major incident occurs, preparation and training are key to responding properly. Under these circumstances, you must:

1. Cordon off and take control of the immediate accident scene or incident site.

2. Care for injured personnel and get medical support to the correct location.
3. Set up secondary control measures and take care of family members, friends and co-workers next.
4. Initiate the incident investigation process.
5. Coordinate and send an initial report to federal, state and local law enforcement and regulatory agencies as required.
6. Prepare an initial response for the news media.

Companies that are not well prepared to respond appropriately can easily damage their reputations. Businesses must be prepared to respond and work with Federal, State and local officials. Always respond to Cal-OSHA/Fed-OSHA when compliance could be an issue because doing so may reduce your exposure to a citation. After the incident, you will be expected to:

1. Manage any potential legal implications in the aftermath of an accident
2. Recover from the incident as quickly as possible and resume normal business operations
3. Report and respond to Cal OSHA/Fed-OSHA as a follow-up measure and maintain any necessary communication
4. Provide counseling services as required

During the aftermath of any major event or incident, things are in disarray and it is a difficult time for all concerned. This is a great time to consider help from outside of your organization; having a list of Safety and Environmental experts, vendors and suppliers on hand, that can help you recover from unexpected problems will shorten your business interruptions to help ease some of the stress.

It is a challenging but rewarding endeavor to seek new ways to make the workplace safer. We all benefit from the products and services that are delivered to our doorsteps, offered on the internet or that we pick up while we are out shopping in the marketplaces that we routinely visit. Always being prepared to respond leaves you in the best position to manage emergency situations, which often occur without warning!

Chapter 4

Safety Incentive Programs – Yes...No...Or Maybe So!

Most people avoid talking a lot about safety incentives because opinions vary so widely depending on an individual's personal or corporate experience. There are many pros and cons in general with employee incentive programs. If you're trying to provide a safety incentive program to encourage higher levels of safety performance and optimum behavior, you have lot of areas to consider, review and decide on before putting a program in place.

Safety Incentive Programs can help an organization reach some fairly aggressive goals. One of the most important aspects of incentive programs is the need to develop a thorough understanding of what the company hopes to accomplish with such a program. Most safety experts agree that being very specific in what the company wants to accomplish is a great starting point.

All incentive programs should be designed and centered on employees. If left up to the management team to design the entire program, it may end up being organized without any true consideration for what your employees care about the most! In other words, it is important to understand what they will get excited about if anything. Your best shot at an incentive program they will pay attention to and want to participate in is having a program that employees help design. If the program is poorly written with the wrong type of awards and recognitions, it will most likely fail.

If the program is expensive and time-consuming it tends to discourage senior managers from supporting it so it is crucial to design one that gets results right away! Some companies have mistakenly viewed their program as being successful when they were actually focused on short-term results. Only to later find that it was short-lived because it was too costly to maintain and did not achieve their long-term goals. An incentive program must show a true return on the investment (ROI).

One of the easiest ways to discover whether or not employees will find the program interesting enough to get excited about participating in is to run a series of tests to see what works and what doesn't! Be aware that not all managers believe that incentives work and some will be reluctant because they believe it becomes more of an expectation by employees, rather than a true incentive. Workforce maturity is an important factor to consider when incentives are in question.

Here are some interesting facts, when thinking about maturity and engaged employees. The key word or phrase being used to describe today's companies who have a better than average bottom line results is (**engaged employees**) and they are usually the difference-makers. Companies with higher than average engaged employees enjoy 275% higher profits, 50% higher sales, 50% higher customer loyalty, 38% above average productivity, 155% higher stock returns!

*This number is huge enough to inspire skepticism, so here's a bit more detail: For 1998-2006, the average cumulative stock return for the S & P 500 was 45.6%. For Fortune's 100 Best Companies to Work for, it was 200.6%. Reference sources: 2013 IntelliSpend Solutions, LLC.

Americans change jobs on an average of 10 times by the time they reach age 40; only 45% of employees were satisfied with their jobs in 2010; job turnover rates remain high since 2009 when it was 16%. In addition, companies say that 85% of their value is wrapped up in talent, knowledge and reputation. In other words, most of their value comes in the form of their human assets. People are generally motivated by many different things throughout their lifetime; it is important to understand your company culture, its maturity level and the behaviors you want to influence, before making final decisions about an incentive process.

The frequency of an award may also change the scope and employee's response to your incentive program. If you wait too long to reward good behaviors, you may lose participation in the program. It has been said that frequent rewards tend to maintain higher levels of interest. If you believe that safety incentives or for that matter any employee incentive program places too much of a cost burden on the company, there are many cost-effective incentives. Some employees see perks like special parking places as an effective incentive or special reward! Be creative and find others.

With the internet as a resource, you can find low cost or no cost ideas that will boost morale, productivity and can easily fit into a well-written safety incentive program. Some companies offer special discounts to their employee's family, giving them free corporate health and fitness memberships as a reward for working safely.

As far as driving the desired behaviors at work; remember that a proper implementation process is critical, never design a program that may be viewed as manipulative and insist on management being absolutely tied to the process of supporting, encouraging and promoting the program once it is adopted.

A <u>Safety Incentive Program</u> should be treated as an important part of an exciting package! For it to be successful, you want it focused on inspiring correct behaviors, designed by both management and the employees. In addition, make sure that it includes specific objectives, remains timely, is constructed well, and properly supported by senior management.

Chapter 5

Safety is a Matter of Leadership Excellence

Safety excellence is deeply rooted in having a philosophy that holds the highest regard for human life. It uses our own fundamental beliefs to rationally justify our decisions based upon what we believe is true and just. Whether it's your personal or business philosophy, the lens that you decide to use when you view the world will almost always be affected by your philosophy. Your philosophy will determine what you value most and it will provide you the courage and wisdom to act on life's most important issues.

Think of this example to illustrate how our philosophy affects our decision-making process. Two groups working for the same corporation but in different cities and providing the same exact services for their clients except for one thing, they experience two distinctly different outcomes with regard to the number of injuries they endure in any given year.

Group 'A' chooses to focus on profits and Group 'B' decides to focus on building relationships. Both groups must make a profit in order to survive, so while Group 'A' spends all of its time doing everything it can to raise their profit margins; which includes charging the most that it possibly can to reach the profit margins they desire. Group 'B' charges a modest amount for their services and struggles but continues to provide quality service and works on building their long-term relationship with key players in the marketplace. Group 'A' maintains a philosophy that as they grow larger they can absorb their injury costs.

Group 'B' on the other hand has a philosophy that holds the highest regard for human safety and realizes the importance of preventing injuries in the first place. They train and encourage their client's employees to focus on safety and health at work and improve their own work lives. Over time, one implodes due to high injury costs and other internal conflicts while the other group reaches its long-term goals of growth and profitability by providing consistent quality, service and obtaining referrals. Same services but very different philosophies!

Let's discuss an entirely different set of circumstances. Worker's compensation premiums are based on several factors; one of which is the experience modification rate (EMR), which is the adjustment of annual premium based on a company's previous loss experience.

Say that Group 'A' has a low (EMR) which lowers their worker's compensation insurance premiums for a given policy year. This occurred because the company incorporated and espoused on the importance of safety throughout their organization. This is a powerful move when done correctly because it sets the tone for better outcomes based upon a philosophy that includes the good of whole organization! It's not based upon selfishness or personal agendas, it's just for the good of the company and therefore it results in lower costs to the company.

On the other hand, Group 'B' maintains a philosophy that supports and reinforces the need to do whatever it takes to get the product or service out the door! They believe their very survival is on the line daily and in order to remain competitive, they must take short cuts, push aside or ignore safety issues and pay little attention to their workforce problems!

This philosophy isn't published or spoken about aloud because that would outwardly draw attention to them. However, everyone in this type of organization understands what the company values most and over time, become part of the problem since they believe they must follow their leadership!

To that end, they experience higher injury rates, suffer greater losses and pay higher worker's compensation premiums because of poor safety performance. What was just described in the two previous stories is a classic leadership failure. This is an example of two real life situations; there are many similar stories that have occurred far too often within the business community. Perhaps it is time to look at our current situations, check for these poor leadership traits and think about how a shift in philosophy could change the safety culture of an organization.

Obviously, the right philosophy affects the entire company from many different business aspects. This book wasn't designed for a select group; it is intended to help anyone, whether they're currently in a leadership role or they hope to be some day.

Using this scenario illustrates how difficult your managerial role can become without the right philosophy guiding you in your leadership journey. There is no work around solution, you must decide to adopt a philosophy that supports safety and live it daily. Safety is a matter of leadership excellence and once we acknowledge that it affects every aspect of business, we can begin to view safety through a different set of lenses.

Another major challenge facing leaders today is the fact that leaders are expected to watch and interpret huge amounts of information very quickly and make decisions with little time to analyze all of the available information. Managers soon become overwhelmed with work and at some point; they seem to only pay attention to their immediate problems. Eventually this lack of focus coupled with a non-existent safety philosophy sets them up for failure. Moreover, when they least expect it, they are caught off guard and that once in a lifetime traumatic accident occurs.

The answer to this dilemma is for today's leaders to have the right philosophy, which ultimately drives their decisions. Safety is not just a management problem, but, more importantly, it is a matter of leadership excellence! We can ill afford to ignore this problem any longer and we must find the courage to embrace and manage it if we are to stay ahead of the economic curve of the 21st century.

What are your responsibilities and what can you do to make an impact on your organization? This seems to be a loaded question because most organizations don't have _'Best-in Class'_ safety programs yet and most employees think managers carry the burden of improving safety. This is far from the truth.

Perhaps the best place to start is with management because what gets noticed in business is almost always cost first! Here is a 3-step approach you can use to begin building a safety culture that works!

Step One. Perform a **safety program and risk assessment** to help you develop a clear understanding of whether you are meeting your safety management and compliance requirements. Once you know this, you can determine the extent of your program management gaps and begin to understand and develop a plan to improve your safety management oversight responsibilities.

Step two. Conduct a physical safety audit. The safety audit is actually designed to identify significant safety compliance gaps. Doing this will help you to determine how to close those gaps most efficiently. Whether or not it's a leadership or management system problem at this point isn't important. The main objective is to close any known deficiencies promptly so that you can reduce your exposure to fines and regulatory problems.

This type of audit should involve a combination of senior management, human resource, operations and safety personnel or the safety committee. The team would then document their findings and take any necessary actions or develop a corrective action plan with the objective of closing out the findings as soon as possible.

Step three. <u>Establish your 'Safety Plan'</u> or what's known in California as the Injury & Illness Plan (IIPP) that was discussed in chapter 3. This will ensure that you are meeting the minimum compliance requirement set by the California Occupational Safety and Health Administration (Cal-OSHA). Cal-OSHA requires eight program elements to be included in the IIPP. These program elements are the basic building blocks of a safety program and they are an important part of building a <u>Best-in-Class</u> safety program! Imagine elevating your current safety program to the next level without these building blocks in place. Here are the eight elements and a short explanation of the role they play.

1. Responsibility – This part of the safety plan identifies a single point of contact that is responsible for updating the plan and ensuring that management's ability to achieve its goals are incorporated into the organizations' mission.
2. Compliance – This section has to do with establishing the methods the organization will take to meet their regulatory obligations.
3. Communication – Without a well-orchestrated communication process, certain aspects of the safety

program may not be understood by employees. A good process keeps them informed and working inside the safety loop so they understand their individual roles.

4. Hazard Assessment – Every organization is responsible for identifying their potential workplace hazards; this section deals with how the organization will meet this responsibility.

5. Accident/Exposure Investigation – All accidents must be investigated and their causes identified and eliminated to prevent recurrences

6. Hazard Correction – Once hazards are identified, a corrective action plan should be devised to prioritize and track hazards until they are corrected and closed out

7. Training and Instruction – Training and instruction prepares employees to perform their jobs safely.

8. Recordkeeping – Organizations are required to maintain documentation that reflects they have a safety program in place that is affective, they must maintain certain required records as proof of regulatory compliance.

Once the basic safety plan you establish is working, you may be able to continue that momentum by adopting some advanced program strategies. These particular improvement strategies prepare the organization to reach the next level and add content and depth to your existing safety program! These strategies should be put in place in order to build a stronger commitment to achieving the organizations' long-term objectives.

Advanced Program Improvement Strategy 1 – <u>A Safety and Health Management System</u> includes your corporate vision, policies and procedures. It also includes things such as management commitment, accountability, employee involvement, feedback, communications and safety performance criteria. There is a huge emphasis on leadership that must be carried out through all levels of the organization. A strong management system ties the overall organization's vision and goals for the future and management's commitment to those individuals who are expected to get the job done!

You must provide details as this paints the picture for employees so they can understand exactly what the company or organization stands for and hopes to achieve in the marketplace. In other words, people need to buy into the vision, support it and feel as though they're a significant part of what it's attempting to accomplish.

It is extremely difficult to hold people accountable unless it is tied to a complete system that's written, published and reviewed regularly. Checks and balances are essential to monitoring our behavior. People have a tendency to slow down or get off track; therefore, the checks and balances are in place to keep us on track and gauge our performance against the goals we have established.

High Achievers want to know where they stand with regard to their individual efforts against some sort of performance standard. Companies that do not have a system for holding people accountable and measuring their performance almost always find that it impedes productivity. As an example, it creates poor relationships, increases tension and problems associated with employee recognition and advancement. When employees do not receive the recognition and reward they deserve they begin to become less productive and fail to complete tasks on time.

Performance standards have an important role to play, however, creating fair standards, monitoring them and reviewing them often can be daunting. To be successful, this essential element should become a part of the overall system for managing people. Employees must be included in developing performance objectives. Employees are more likely to engage and commit when they have had an opportunity to add their perspective. This builds trust, also most people want to feel included and know they have had a say in shaping their own future.

At Miller Brewing Company, we spent hundreds of hours designing specific campaigns to excite and reward certain milestones on our way to achieving our agreed upon goals and objectives. The key phrase there was *agreed upon goals* between the managers and their employees, which were aligned with supporting the corporate, and brewery missions. Corporate goals were communicated routinely throughout the workday.

We had television monitors running our greatest commercials, superstar race teams promoting our company, Miller employees featured in our commercials, corporate executives reaching out to speak with employees and a host of specialized reward programs to ensure that we continued to recognize and acknowledge high performance! This inspired everyone to become high achievers for the good of our corporate identity.

These actions made everyone feel proud to be connected with Miller Brewing Company and as far as safety performance was concerned, it became a part of each person's daily agenda. The most challenging issue for companies who reach for the higher levels of performance in this noteworthy area is sustainability. Sustainability means hanging on to long-term high performance results that last for 5-10-15 years or more!

Advanced Program Improvement Strategy 2 – The orientation, education and training program portion of any safety process is extremely important. Your employee orientation sets the tone and it provides employees with specific information they can refer to when they have questions about the company's core values or important standing policies.

The orientation is the best place and time to bring all of the company's key resources in to one place to discuss roles and responsibilities and describe how each person's role fits into the organization's mission. To get the best results the orientation process should include senior management and each department supervisor whenever possible.

Some of the best-known programs have a tiered approach where executives speak to all employees about their particular take on how safety, health and the environment affect their business and its reputation. They clearly explain their roles in preparing for a future that's bright and inclusive for all; an extremely important part of their message is publicizing their scoreboard.

You want your employees to understand how the company evaluates their performance and what the key measurement metrics are before they begin their careers with a company that has 'Best-in-Class' aspirations! Because safety is a strategic objective, it should be discussed during all formal training and education classes as well as during the employee orientations.

Supervisor safety leadership and management training should be provided to anyone who holds a position of authority and is responsible for employee safety. A basic understanding of the corporate policy and procedures for administering and managing safety, health and the environment is paramount to the success of an organization. Remember that standards must be understood and applied equally to outside vendors and suppliers as much as they are required for employees. This is a huge challenge and stands at the heart of every *'Best' in Class'* organization. The safety posture of an organization is set by its leaders and employees are always watching what management does with regard to safety.

Vendors and suppliers who provide services and enter the company's premises must be prepared to comply with an organization's safety rules while they are on-site. Doing this establishes a consistent message throughout the company that safety applies to everyone.

Successful companies reduce their risk and improve safety by focusing energies on building great relationships and providing top-notch training! If first level supervisors are respected and they buy into the necessity of training; their employees will see the need as well.

There was a company located in Southern California that was spending about $30,000 a quarter for their emergency response training. This emergency response team was expected to respond to emergencies that involved ammonia hazards as well as common safety and medical emergencies. However, the team became complacent and began to take advantage of the system and their positions on this highly specialized response team.

For some time the trainer had been ineffective due to a lack of employee motivation and participation and felt as though his relationship with the company was in jeopardy. When the problem first arose he decided to bring it to managements' attention, but it wasn't resolved at that time. He continued to work and provide the training until the problem exploded into a confrontation between him and the attendees. The trainer was caught in a precarious position with the company and union. The trainer felt as though he could no longer place people at risk or waste their resources and once again brought the problem to management.

The senior manager attended some training sessions over the next several months and little by little, they raised the membership requirements. A few of the smarter individuals decided to remove themselves from the team because they understood where things were headed. Eventually the time came to publicly address the team while it was in session.

The most critical training was conducted the first week of each quarter. On the first day of the second quarter, the manager walked in and sat in the back of the classroom, mostly unnoticed! A couple employees happened to notice later that he was in the back of the room.

As the instructor reviewed the agenda and began teaching, he faced unprecedented challenges with class participation. Now this training was held onsite in a state of the art facility, far removed from the main campus. This minimized interruptions during training sessions. When the team was in training, they received a gourmet breakfast and lunch daily.

Snacks were delivered by the site's cafeteria that included coffee, water, juice, soda and anything your sweet tooth desired. They even took into consideration special diets and quite often-other specialized food requests.

The training facility was equipped with bathrooms, separate break rooms, telephone, fax capabilities, private parking and all of the audio-visual aid equipment that one could dream necessary to support their training. Two different fire departments and their hazardous materials teams periodically attended these training sessions and they also participated in joint exercises to help these teams practice for real life emergencies.

This emergency response program was designed to attract bright, dedicated men and women who were physically fit and who could perform under some unique conditions. In return, they expected the team to represent the company and be stellar employees. The team received public recognition within the community as well as throughout the company. They enjoyed special privileges, and other benefits that went along with being a member of this team.

The manager sat in the back of the class for a while and observed the disruptive nature of some of the team members. Here is what he witnessed; some women were knitting in class, some were talking, and a few men were being extremely rude by making odd comments while the trainer was attempting to go over the days' agenda. The manager was embarrassed and at that point, it was obvious that some extreme measures were needed. At that moment, he rose up from his seat and most of the class was caught off guard.

The manager asked the following question, who can afford to pay back the cost of this training and the equipment that we have provided you by the end of this fiscal quarter? Not one person raised their hand or responded. Then he asked another question; why would a company spend such a hefty amount of money on this type of training? One person was brave enough to stand and say, "Because we may actually be faced with the real possibility of having to save a life". Kudos to him….

It did not matter whether they were there for the perks or the challenge. He told them if they were here for the perks and not for preparation of duty to please be courteous enough to remove themselves from the team immediately. For if, they had to waste precious time on this topic again that he would ensure they were removed.

Training and education is important and is often a critical part of any safety process. This particular manager made it a point that day and showed that leaders are often called upon to take immediate action. This illustrates how important training is to an organization and training should never be taken for granted. Training and education provides the first and best chance to set people up for success, no organization can afford to waste precious training resources!

Advanced Program Improvement Strategy 3 – A Behavior Based Safety Program is concerned with the manner in which people carry out their specific daily and operational tasks. It is essential to understanding what makes people click and do the things they do...in other words; the principles and laws that govern our behavior as studied and explained by Behaviorists!

Large corporations have used Behavior-based safety approaches to improve organizational safety for years now. Most formal programs require a significant investment in time and money and most small to mid-sized companies do not usually have the budgets, staffing, specialists and other resources necessary to put a 'Behavior Based Safety Program' in place.

Behavior based safety is the process of collecting information (observable data) on how employees are conducting their tasks. This produces information that when analyzed using the scientific method should give way to understanding some possible approaches to improving safety. The scientific method involves:

• Identifying the problem or opportunities for improvement
• Collecting the information or data
• Developing a hypothesis
• Testing the hypothesis
• Drawing some conclusions
• Applying your findings to test the hypothesis under other similar conditions and or environments

Small and mid-sized companies can still use some simple tactics used in behavior-based safety with good results by taking an informal approach and following the steps listed below:

a) Get to know the employees first by showing that you're curious about what they do
b) Show interests in their role and responsibility within the company
c) Learn how and why they do what they do to complete their tasks
d) Ask permission to observe and explain why you want to know about their jobs
e) Ensure they understand the value they bring to the company
f) At the right time, state your intentions and let employees know what the goal of behavior based safety observations is, ask for their help and explain how you can use it to improve their work areas

With this approach, you can see first-hand what is causing problems for your employees. This way you will be in a much better position to evaluate and solve their problems. Never offer recommendations for improvement without communicating with all stakeholders first. It's important to do this throughout the process and well before any short-term or strategic changes are implemented.

You will get the best results by including your employee's suggestions and feedback into the process. This way you will be sure that employees will follow new procedures and processes or use new equipment properly. The test phase should provide accurate information about whether the desired outcomes that you expected are actually occurring once they have been implemented.

In most cases, the behavior-based safety process is best managed by a group that has an immediate concern or stake in the program. A Safety Committee or Mastermind Group is usually best suited to do this if they are authorized to make immediate changes.

The committee or mastermind group must be able to respond and take action on issues that affect production, interrupt shipping, cause quality control issues or other similar problems.

The reason for using Safety Committees or Mastermind Groups is to ensure the sustainability of the program. Your greatest achievements and successes can be lost over time if the program is not maintained. In companies that have unions it is important to ensure, they are on board as well. Most unions should have a stake in a behavior-based safety process since it focuses on making workplace improvements.

The Safety Committee or Mastermind's continuing role is to stay abreast on training, observe the work processes, collect data and keep management informed on the status of the program. Keeping management updated usually generates continuing support on their part if there is tangible proof the program is working.

If the goal is to reach _'Best-in-Class'_, it will be necessary to consider integrating some amount of behavior-based safety into your overall program at some point. Once the other basic safety program elements are in place and you've had time to evaluate your programs effectiveness; you should be able to generate support for a behavior-based strategy program to round out your program in an attempt to reach _Best-in Class_!

The cost and time associated with implementing this type of process is usually well worth the effort. However, you will want to ensure that your basic safety program is working well before attempting this type of advanced program strategy.

**Advanced Program Improvement Strategy 4** Compliance Management – Most small business owners believe that compliance is a necessary evil; now that's not to say they don't see the benefits of safety or regulations, but, many do feel as though we've done a poor job of effectively writing regulations. The California worker's compensation system is probably viewed by most business people as inadequate and technically broken. Safety and health regulations are often complicated and so they are easily misunderstood by most people. In many cases, the requirements are not feasible to implement, they are constantly challenged or are often ignored.

Compliance Officers often have their own difficulty explaining regulations so they use special clauses and letters of interpretation to escape justifying poorly written regulations. If they have difficulty citing an area they believe is a safety or health concern they fall back on these clauses which seems unfair to many business owners. Business owners are in a tough position because most of them realize the importance of safety and they understand that standards are needed, but it frustrates them to see companies that do not meet workplace safety and health standards.

In the past companies relied heavily on off-the-shelf safety programs because they sought quick fixes to meet their regulatory obligations. Most of them failed to realize their programs did not meet the intent of regulations because they were not customized for their operations. Some were issued citations for non-compliance; that occurs less frequently in today's environment but the possibility remains if they do not customize their off-the-shelf safety programs.

This is why the most effective safety consultants seek to understand exactly what the safety and health management system of a company looks like before proceeding to offer their assistance. Basic compliance is a matter of assessing the risk as it relates to current and applicable standards as well as the common sense issues that occur daily. Checklists are used to help keep an eye on compliance and they serve as a great tool to document your internal safety inspections. There are some basic safety checklists that apply to all businesses and then others that would apply to your specific operations.

Advanced Program Improvement Strategy 5 – Industrial Hygiene and Occupational Health is an area that smaller companies usually learn about as a result of an employee complaint or health condition. As an example, if a company has machines and equipment that create high noise levels, they may be required to have a Hearing Conservation Program. If there are chemical exposures, they may be required to conduct chemical sampling and implement an Industrial Hygiene Program.

If there is a need for an industrial hygiene program because there is possible chemical, hazardous materials, explosives or other exotic materials present at work; there are three actions that a business may take to meet their basic industrial hygiene, health and environmental compliance responsibilities and reduce or eliminate the risk of being cited by authorities.

1. First, contact a reputable **occupational health and medical facility** and request that a physician's worksite visit to your facility or filed site to observe what your employees do at work. The medical facility visit should also include their Senior Administrator since this is usually an important contact person. Doing this will familiarize your medical facility with your operations. Company representatives should also visit the medical facility and receive an orientation and overview of their capabilities as well.

2. Then call some **industrial hygiene firms** to get quotes for conducting a baseline survey of your facility. This type of report will identify if there are any chemical exposures and whether or not you need to implement a formal industrial hygiene program. It is best to insist on having a 'Certified Industrial Hygienist' that oversees and manages the sampling project. Any baseline survey should be kept on file for your reference.

3. Your next call should be to a reputable **environmental company** to get an assessment in order to ensure that you meet the basic environmental compliance requirements for your industry, operations or locale.

In each of the situations you should acquire multiple quotes and compare them line by line in order to understand and properly compare their services and pricing. Have them provide you a reference for each service that is required by a regulatory agency. Never purchase a service plan right away without completing your own due diligence. You may also want to contact one of your industry associations.

Industry associations may offer some assistance as part of their membership plan. They are usually well worth the cost, especially for smaller companies who do not have the staff or time to research the requirements. Industry associations may also offer discounted education and training programs as well.

Advanced Program Improvement Strategy 6 – Worker's Compensation Case Management has more to do with the incident reporting process, analysis and managing the care management of your employees as a result of an accident. Large companies are typically self-insured, meaning they're licensed by the state to provide their employees injury insurance and they have the staff and expertise or third party administrators that are licensed to provide for the care, coverage and management of their injured employees.

For most small companies the best way to control your claims cost is by working as close as possible with your carrier. Employers who are large enough usually choose to self-insure as a method of achieving greater control over their claims management process. The success of a workers' compensation self-insurance program will depend upon the effectiveness of their loss control activities and claims supervision.

Most self-insured employer's contract with third-party administrators to manage their claims and a few will employ their own claims management departments. The truth is that the process is expensive, complex and the financial requirements are extensive to self-insure. If you would like some additional information visit: (http://www.dir.ca.gov/osip/apprequirements.htm).

To receive self-insured status employers must apply and meet some specific financial criteria. Once approved by the Director of Industrial Relations they are listed with the state as self-insured. For more online information concerning the application process visit: (http://www.dir.ca.gov/osip/apprequirements.htm)

Most start-up companies begin coverage with the California State Compensation Insurance Fund because there is no loss history to base their premiums on; therefore, they are viewed as a high-risk client. Insurance Brokers usually place high-risk clients with the State Compensation Insurance Fund when they are just starting out, have a poor safety record or high loss history.

This is because traditional insurance is either too expensive or they won't cover the company. Even when placed with the insurance fund the cost is usually much higher than traditional insurance markets. It's just not good for business or a company's reputation to pay these higher premiums or to be on record as having a poor loss history.

The best possible scenario is to maintain a low risk posture; especially since there is a three-year loss history that affects the base worker's compensation premium rate that your company will pay. In today's competitive environment, it's extremely important to keep all of your costs as low as possible!

The California State Compensation Insurance Fund is the largest and most expensive form of worker's compensation in the country. It is often the only choice for companies who have poor safety records, high injury rates and losses! Why, because traditional insurance companies don't want to underwrite such high risks, they're also in business to make a profit. This leads us to the importance of reaching for *'Best-in-Class'* safety status! A poor safety program results in huge financial losses seen in (both direct and indirect costs). These higher insurance rates are paid out on each payroll and throughout the claims process once an injury claim has been filed.

The cost of worker's compensation can be the difference-maker. In California there have many companies who have closed their business or reduced their payrolls to absorb these higher costs. For many business owners obtaining the lowest possible workers' compensation premiums was the single most important factor that allowed them to continue their operations.

Advanced Program Improvement Strategy 7 – Emergency Response/Crisis Management Your response plan must be customized to address the specific needs of your facility. Each location is unique and has its own set of potential risks and special hazards. One of the basic considerations is fire prevention and facility protection.

The local Fire Marshall sets building occupancy levels and inspection requirements. Both Federal and State safety regulations require evacuation procedures and maps that indicate how to egress the building safely in an emergency. The overall plan must include placement of fire extinguishers, sprinkler management systems, whatever is required by the code. Practice drills should be conducted to ensure that employees can actually escape safely and that management has the ability to account for their employees.

One of the worst potential tragedies is sending a fire fighter into a burning or collapsing building to look for someone who isn't inside the building. Unfortunately, these sorts of things have occurred to often; employees must be trained about the serious nature of emergencies. The primary emphasis for any emergency action plan is having the capability of preventing deaths and serious bodily injury.

An Emergency Response and Crisis Management Plan is another critical set of guidelines that can make or break a company within as little as 48 hours. The management team is sure to fail if they have not practiced and don't understand the nature of responding to the public, media, family and friends of employees in a timely manner once a crisis has occurred.

The need for a crisis management system that is designed to work is one of the greatest attributes of a 'Best-in-Class' organization. A well-thought out plan reflects the highest regard for human life under the toughest conditions imaginable.

Advanced Program Improvement Strategy 8 – The Five Essential Daily Management Actions that must become part of your routine:

1. Automate as much as possible because this will minimize wasted time and maximize your supervisors' ability to be effective in leading their departments.

2. Remove roadblocks that prevent your senior officials from getting out and spending time with employees. They must spend that critical time talking with and engaging employees, really asking about their problems, and listening to understand how they can eliminate roadblocks that prevent them from achieving their goals. These impediments may seem small, however, they are huge if it frustrates employees, forces them to slow production, impacts quality, tires them out before they should be or increases their risk of injury!

3. Seek employee's suggestions by asking for their suggestions about improving the workplace. Once you've been entrusted with this information, act on it or you may risk losing their trust. When changes are made that were based on employee feedback, you should publicize that to show the company is listening.

   Some suggestions may require personal recognition between the supervisor and their employees. Perhaps some recognition may be warranted and be appropriate for public recognition. Before you recognize an employee publicly, you should get their permission. No matter what you do, you should thank employees for their suggestions and for providing you feedback.

4. Law Firms should be used with caution and we highly recommend spreading your litigated claims between multiple law firms. If you are currently sending all of your litigated claims to one law firm you should stop that practice immediately! You may

want to get additional referrals from some of your business colleagues. No matter what, you should always spread your caseload among several law firms.

5. <u>Use law firms that specialize in worker's compensation law</u>, do this as a precaution and let each law firm know that you respect their professionalism but that you have high standards and expectations. In other words, they must continually earn your business.

Safety is a matter of leadership excellence; nothing great can be sustained without ensuring the right philosophy is in place. A leader's safety philosophy is at work throughout the entire organization; people buy into the philosophy and believe it is important when it is continually communicated, demonstrated and most of all; they're shown serious appreciation for their commitment to safety excellence!

Chapter 6

A Brief Look at The Worker's Compensation System

California has the largest state compensation fund and the second costliest plan. The original plan was designed to provide wage compensation, medical and death benefits. These benefits are paid to employees who are injured or killed during the course of their employment. In exchange for these benefits, employers are not supposed to be sued under our tort law. Workers' compensation is the sole remedy for injured employees; but there are exceptions as governed by state law. <u>Under the following conditions, a company may not liable for an employee's injury claim.</u>

- injuries caused by intoxication, drugs, illicit or illegal drugs
- any self-inflicted injuries or as a result of a fight that was started by the employee
- horseplay
- claims of injuries by an independent contractor (the burden of proof for independent contractors weighs heavily of the employer, make sure you have a rock-solid process for employing and verifying their status)

- violation of company policy
- injuries during a felony-related incident
- any off the job injury or illness related to a non-work condition
- post-termination claims

Here are some situations where an employer could possibly be sued:

- employer is found to have operated recklessly outside of the law
- committed intentional actions that caused an injury to occur to an employee
- auto accidents under certain conditions (employees have successfully sued outside third parties who were at fault) or (where their injuries were caused due to faulty or defective equipment)

The type of strategy that a company adopts to manage their workers' compensation claims is a huge determinant in controlling cost. Each state has its own set of procedural rules, rates paid to employees, procedural time constraints for mandatory actions and waiting periods that may also affect cost containment efforts.

One way to stay informed is to check the internet frequently and research the basic rules and requirements for your state by searching the web for information under workers' compensation or state labor law. Insurance industry and trade associations are a great resource for training and information. Even when business owners have injury reporting procedures in place, employees often delay reporting injuries or they do not report them at all until some other problem surfaces.

When faced with a late report of injury, supervisors and managers may become frustrated even though the employee is technically covered by the law. This is the time to exercise patience because it often pays off in the end. Once notified of an injury claim an accident investigation should be started immediately.

Companies must remain objective and be careful to follow the rules for managing worker's compensation. Small business owners may become directly involved in an injury claim and at times that has proven to be troublesome so be careful. For companies who manage their own claims in-house, the investigation process will often reveal conflicting information, different from what is uncovered by the insurance carrier. This will sometimes affect their thoughts about the employee or their claim.

The best advice is usually to let your insurance carrier provide the experts to manage your claims. This team of experts will protect your interest. No matter what, you should always keep an eye on your cases and hold quarterly claims review meetings with your insurance carrier.

Once an employer ceases to have employees or is out of business, worker's compensation insurance can be dropped but there are four types of injury claims that an organization must understand. These claims must be substantiated by a physician's medical findings, which are then used as the basis for an employee's claim.

1, Temporary Disability
2, Medical Treatment
3, Permanent Partial
4, Vocational Rehabilitation

There is a serious and willful misconduct clause of which employers should also be careful. A legal finding against an employer could increase employee benefits by as much as fifty percent (50%). A death benefit would apply in those cases as if they were upheld. Many state legislators, judges, attorneys and business owners agree there are flaws in the current worker's compensation system. The system is further complicated by many other factors such as:

- Misunderstanding of the benefits and timeliness of distribution
- Maltreatment of employees by management or medical treatment facilities
- Poor communication
- Claims processing bottlenecks
- The accident/incident investigation process
- Third party insurance claims administration
- False claim filings and active investigations
- The litigation process and tie-ups in court
- Differences in court jurisdictions and rulings from county to county
- Law firms and legal services
- Disappearance of litigants or not showing up for court dates (in California, many employees leave the state and go to Mexico where other family members live)
- Third party participation and recovery of expenditures

It is very important to make sure that all employees understand their worker's compensation benefits before a situation arises out of employment. When an accident occurs, human resource teams should sit down with injured employees and clearly communicate what the system pays and keep them informed about the status of their claim(s). Employees should feel there is no hidden agenda when they are sent for medical evaluations or treatment.

There have been cases reported where medical clinics treated injured employees poorly by leaving them unattended and waiting too long to see a physician. Employees will also complain if they are not provided clear information about their medical condition. It is important to ensure that employees understand their diagnosis and have all of their questions answered to their satisfaction.

The best way to reduce bottlenecks is to ensure that your employees complete their paperwork in a timely manner. Failure to meet reporting periods may slow down processing of a claim for payment or even disrupt their medical appointments.

Chapter 7

The Reach for 'Best-in-Class' Safety Excellence

So why doesn't every company or organization achieve a high level of safety success? Why aren't there more companies experiencing what's commonly referred to as *'Best-in-Class'* performance? Because it takes the right type of philosophy, technique, patience and creativity to make systemic changes that become sustainable.

In the long run, if they do not to reach the level of success that others achieve they give up and begin to pursue profits. Eventually things falter and times get tough, they experience cost overruns, suffer quality control issues and ultimately customer service problems appear. Eventually the entire organization begins to feel the pressure of not achieving their objectives.

Despite the size of the company, corporate philosophy influences the overriding goals and objectives of an organization. If reaching *'Best-in-Class'* status is a strategic goal then it is possible to achieve it. If it is not a strategic goal, most organizations eventually fall prey to focusing their efforts on profits rather than attempting to become a *'Best-in-Class'* organization.

The Armed Forces of the United States is a great example of a typical *'Best-in-Class'* organization. Military leaders understand what they do can only be accomplished on the world stage if their philosophy is transparent and understood throughout their organization.

These leaders understand the importance of achieving organizational excellence by paying attention to and achieving *'Best-in-Class'* safety performance! Safety has been the cornerstone of military success for many years. The goal of pursuing *'Best-in-Class'* status should be the standard of every employer. The military seeks to do this on a daily basis because lives depend on it…should it be any different in corporate America?

The Armed Forces understands their success depends upon the quality of training they provide for their troops. This is something our military figured out many years ago. Training has become a major factor in reaching *'Best-in-Class'* status, training requires practice!

Vince Lombardi was one of the greatest football coaches in history. He built his winning teams on hard work! Vince Lombardi Jr. captured the essence of his father's philosophy in the book he wrote titled; "The Lombardi Rules". In it, he wrote that his father placed great emphasis on practice and his father was often overheard yelling, "Run it again".

He said that his father and his assistants would run the same play over and over again until the players themselves were yelling "Run it again"! This continued until they ran it perfectly each time. That is why the military places such great emphasis on training, because they also understand that training forces you to practice and practice eventually leads to excellence and it eventually becomes the way you operate!

We could also take a lesson from our military and utilize the power of cross training. Cross training has been perfected by our military because they often find themselves in some very difficult and life threatening situations, where one person may be called upon to fill another soldier's shoes with only a moments' notice! In the private sector, it could provide an organization the flexibility they need to fill an open position very quickly!

Six more powerful strategies to help reach 'Best-in-Class'

1. Install executive checks and balances - Create a method that allows the company to validate and verify the system is operating within the parameters that have been established. The goal is ensure consistent application of the rules throughout the organization. The executive's role is staying tuned in to the employees as much as possible. To do this well, executives should spend time with their employee's trying to learn whether the culture they envisioned is taking hold. The corporate vision should remain consistent with the public's perception of the company.

2. Invest in the infrastructure- Just as technology has allowed us to remain competitive, people can only remain competitive if companies continually invest in their facilities and infrastructures. The best way to accomplish this is to consider all of the possible situations that could impact employee safety and performance. Whenever major changes are being considered, think about what type of impact new equipment, processes and systems may have on employee safety.

   The work environment should be designed to protect employees against workplace hazards and to allow employees to maximize their skill-sets for their work environment.

3. Focus on supervisor's role in relationship building - The importance of human relations and how it affects workplace dynamics is often overlooked. The first lev-

el supervisor is your best source of employee information and performance because they spend the greatest
amount of time with employees.

Supervisors have the best opportunity to model the
behavior and the culture that you desire. Supervisors
understand who does the work but they don't necessarily look at how their relationships affect productivity,
quality, creativity, problem solving or the desire to go
the extra mile. When an important goal is in reach or
the extra help is necessary to meet a specific objective, employees will participate or volunteer if there
is a strong commitment to the company. That commitment is reinforced by their relationships with their
supervisor.

This is the power and influence that immediate supervisors can have by building the right type of relationships with their teams. The best way to accomplish
this is to provide safety leadership training. Supervisors who have attended formal leadership education
and training do a better job of leading and building
relationships with their employees.

4. Consider a Benchmarking Strategy – This will give
you an external look into what others in your industry may be doing or attempting to do to improve their
work processes. Benchmarking could lead to discov-
ering industry trends that are actually working! Failing to benchmark is a missed opportunity to identify
shortcomings and create improvement strategies. This
may hold a company back from achieving their next
big breakthrough! Benchmarking can provide you
with insight on every aspect of business as well as
providing you the foundation to make good business
decisions.

5. Raise Organizational Skill Levels - The ability to develop your existing talent and resources is smart busi-

ness simply because it is less costly to recruit from within an organization whenever possible. If employee training and skill development was offered to employees as a reward and it enhanced their personal development and growth, it could provide an alternative solution for filling temporary or permanent positions.

People are often left with limited options for job growth and little time if any for developing additional skill sets that could increase their options. Another advantage of offering additional education and skill development training is that it breeds loyalty. Yes, it is also true that you could offer these classes and people might leave and find other opportunities, but that occurs less often than many people think.

6. What is in it for them (WIIFT)-By the time your organization reaches this point, it has probably developed and embraced a champion's mindset! Once a company makes the intentional decision to establish a caring mindset for their employees, their employees usually, reciprocate by deciding their company deserves to win! When employees decide they want to win, they give their best to ensure success. It does take a lot of up front work to establish the right channels of communication.

There are many ways to make sure that your employees understand they are part of a bigger plan and they have a right to expect good things for themselves when they pull together as a team. Publicizing everything you offer them is a way to entice your employees and encourage them to participate in the safety process. It creates a win-win scenario. Timely communication eliminates fear and prevents rumors while building trust. Quite frankly, it clears the way and allows the team to focus on their objectives.

The great minds of our younger generations are being very selective about where they want to work these days. As the demands grow your ability to maintain and recruit talent will be tested and forever challenged. These particular strategies will help to ensure that your company or business remains attractive and properly positioned on the road to safety excellence!

Chapter 8

Behaviors are Shaped at Home and Influenced at Work

As humans, we develop a sense of self and we model our parents or others who care for us during our developmental years. Our behaviors are shaped at home first and then influenced at work due to the amount of time we spend in our work environments. Since we spend so much of our waking moments at work, we become heavily influenced by our organizational culture, subcultures and the work habits we form within the organization. Our fellow associates and team members influence our behaviors as well.

The people we associate with at work come from a variety of backgrounds and they have their own beliefs, values and life events that have shaped their behaviors. The mixing of our cultures, our various behaviors and the conversations that we have with each other weighs heavily on how we respond to various workplace conditions. Leaders should recognize these differences and try to leverage, shape and influence workplace behaviors both at home and at work.

For example, if an employee is required to wear personal protective equipment (PPE) at work in order to perform their job safely. It is reasonable to expect that in time that same employee may see the benefit of wearing personal protective gear while working at home as well.

Managers should encourage the positive behaviors and discourage the negative behaviors that may be affecting job performance. Here is another example; let's say that we have a skilled performer at work who finds it difficult get to work on time, there may be an opportunity to reinforce the desired behavior of showing up on time by encouraging them to make and keep their appointments and commitments at home. Influencing a small behavioral change over time may help the team overcome problems caused by the individual who shows up late for work.

Chapter 9

The Intentional Risk Management Model

Risk management can be explained as the identification, assessment and strategies used to control or mitigate the adverse effects of risk on an organization. In other words, risk management is the organized efforts taken by an organization's management structure to reduce the potential for loss. The intentional risk management model must be supported and approved by senior management. Doing this signals their support of the process and places the model at the center of their strategic planning conversations.

It provides a model that managers throughout the organization can use to support risk management as a strategic objective. This is something they will be held accountable for so it illustrates and reinforces taking any immediate and necessary actions to improve safety. Management teams will often decide on a specific course of action and then delay it because there is lack of support within the organization. By having managers personally commit and adhere to the principles of the intentional risk management model, many of these unnecessary roadblocks are removed.

Senior management must be involved at the executive level in order to establish priorities and time frames for taking corrective actions. They assign the responsible parties and hold their managers accountable for closure. The intentional risk management model is the glue that holds it all together and there are systems that are set up to track the team's progress until each corrective action is completed. This is the basis of Intentional Risk Management; identify, record, correct, and communicate in a timely manner throughout the organization.

The Intentional Risk Management Model establishes an intentional methodology that holds its management teams accountable for setting up, communicating and integrating a process for making systemic changes with regard to the five areas listed below:

1. Organizational Vision/Mission
2. People/Processes
3. Physical Environments
4. Specific Goals
5. Overall Objective

The model is based on a specific philosophy, mindset and principles that will encourage leaders to consider risk in every major decision they make. Since intentional risk management in part is a philosophy and a developed mindset, it must be used in conjunction with a guide or systematic approach. Management performance reviews are also incorporated and used to encourage and measure participation and support by managers.

The **E.S.A. Pyramid** is comprised of three major components, which illustrate how safety excellence is reached and sustained. The E.S.A. Pyramid supports the 'Intentional Risk Management Model' in a very practical manner. From experience, we know that performing at the highest levels requires individuals who have the desire to continually improve and achieve more than average.

**Consider the following illustration
about the E.S.A. Pyramid...**

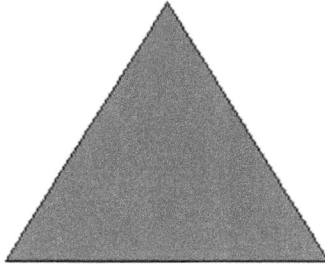

The base of the pyramid is **Education,** on the left is **Strategy** and on the right is **Action.** Each component is necessary to hold the structure together. Once the management team sees the need and benefit of education, they should easily draw the conclusion they need a plan or strategy to support education and training within the organization.

Once a strategy has been developed, the next logical step is to take action to complete the pyramid, doing this should encourage teams to become part of a learning organization as they become part of the learning process.

At that point, we create the momentum necessary for it to become easier to reach the organization's goals. Consider this, if a company fails to invest in a formal training program it will be very difficult for them to remain competitive for very long.

The E.S.A. Pyramid focuses managers on providing a systematic approach to educating their employees, supporting interactive strategic planning and deploying the resources to take corrective actions in a timely manner. This inspires employee participation and support for the safety process!

Education-Strategy-Action enhances safety throughout the organization and it supports needed changes, allows employee participation and gets the results that make a difference in their work environment. The key to education is that it provides a foundation of information that an organization needs to do its job properly. The learning process changes the way people think about their roles as leaders and when organizational leaders learn to connect emotionally and empathize with their employees it builds confidence and loyalty.

Likability and character is another important factor for any leader and it helps answer the employee's question about what's in it for them. Employees need to know what must be accomplished in order to expect them to become engaged in changing the organization.

Good leaders are able to show employees' how they make it possible to compete in today's market and when they are invited into the process of change it builds confidence and trust.

It is extremely important that employees see the possibilities for new opportunities. In many small and mid-sized companies, opportunities aren't discussed or publicly spoken about and this can affect the ability to retain good people. There are many ways to build and promote opportunities within an organization; this can be accomplished through promotions, special assignments, job enrichment, technical training and much more!

Recognizing personal achievement both on and off the job can be a positive force in shaping workplace behavior. Employees and staff must be allowed to have fun at work! People need to know there's more to work than just the hustle and bustle of meeting their deadlines. A workplace where people are free to have some fun creates an atmosphere of belonging, which in turn generates a higher level of commitment. When people become proud of where they work, that alone stimulates discussion and they unknowingly advertise for the company. The **Intentional Risk Management Model** and the **E.S.A. Pyramid** is the gateway for reaching employees on a deeper level. The question of why employees would be willing to give so much of their time, creativity, energy, enthusiasm and effort is answered when an organization consistently values their employees. When people feel valued they give more of themselves. They look for new ways to solve their own problems; some even go as far as to spearhead their own teams, looking for solutions to problems that plague the company.

When employees are free to experiment with new ideas and receive support from senior management, anything is possible! These ideas may require design, technical changes or other resources; the result is often groundbreaking. This gives the team room for discovery, developing formal processes, new policies, programs or ideas that help the organization achieve its goals. Creating this type of work environment can improve safety, quality and a host of other areas within the organization.

The intentional risk management model is supported by the **E.S.A. Pyramid**. In fact, this model could be adapted to change the course of business for any organization in a multitude of ways and is not restricted to the safety improvement process! The ESA pyramid takes into consideration that to be successful at institutionalizing major changes within an organization; the organization itself must be aware that education and continual learning is essential to their growth and prosperity. It's further understood that once the education process begins to take hold or is improved, that any resulting action should be based upon a specific strategy designed to reach the goals and objectives of the organization.

If the goal is to impact organizational change that produces greater results which are sustainable; then we should rely on the intentional risk management model and focus our attention on achieving 'Best-in-Class' for ourselves and our companies.

Chapter 10

Taking the Pain Out of Work!

As a leader, one of the greatest compliments you could receive is to have an immediate and positive impact on an employee, a department, a business unit or an entire organization! The best way to do that is to find out what is not working well and look for new solutions. When you solve organizational problems, senior management tends to listen to your suggestions more intently. Take a look a few key safety lessons that companies have used to get noticed, make an immediate impact and lower their costs. You may find use for these as well.

1. Immediately following any significant accident, if you shut down your operations, you immediately raise the level of importance that you place on safety throughout your entire organization. We did this at Miller Brewing Company and it had an enormous impact on the brewery's safety program. Taking this strong an action gets immediate attention and everyone realizes the impact.

2. Investigate every accident/incident immediately or as soon as feasible. There have been many cases where a seemingly minor accident was not properly investigated because of time or limited resources. The main purpose of an accident investigation is to gather the

pertinent facts and identify how to prevent similar occurrences in the future. However, a proper accident investigation will likely help your insurance provider manage the claims process more effectively.

3. Make sure your occupational medical facility has 24-hour accessibility if at all possible. It is especially important to ensure they are open during your particular business hours. Employees are always concerned about their medical treatment and the quality and care provided. So it is very important to spend the time necessary to ensure you have a good clinic that your employees will have faith in and not question or fight every decision the doctors make. Encourage physicians to visit your facility to evaluate the workplace. You want the clinic's attending physicians to see your work environment so they are familiar with your employee's job tasks. Employees will get a sense that your clinic's staff has seen the work environment, which clears a path for better communication between all parties.

4. Get help from reputable Safety and Health Specialists to work on specific programs or policies that you don't have the expertise to accomplish. However, remember to be very selective when contracting safety and health professionals.

5. Develop a mandatory modified duty program, identify modified job duties and establish performance standards for each job classification before you need them. If you don't have a written modified duty plan and you neglect developing specific modified jobs then you may experience legal problems during the claims process.

6. Develop a written drug and alcohol program with a mandatory termination policy in place. Doing this may prevent or reduce the risk of impairment related accidents.

Final Thoughts

Ultimately, the road to excellence in safety must be carved by leaders whose philosophy is built on principles that hold human life indescribably important and valuable. At some time in your career people say that you should not take your job so seriously. When you hear this, listen, it may be an indication that you are on the verge of alienating people or losing your focus. The strategies, recommendations and thoughts captured in this book will help solve many of the routine safety issues that we face throughout our professional and management careers.

I hope that you will find enough value in this book to use it as a desk reference or have it become required reading for your managers. It is also intended to generate new conversations about how leaders should view the workplace in order to reach the higher echelons of achievement, excellence and overall success.

There are some short cuts to building safety into an organization. Some of these ideas are new and some are tried and tested short cuts that have been proven to work. Most will save time and prevent you from wasting precious resources or reinventing the wheel.

It is also worth having weekly conversations with your employees and staff about the relevance of safety to the organization, your employees and the corporate mission. This sort of communication is a reminder of the importance that is placed on workplace safety and may reduce or eliminate the mistakes that have caused problems for other organizations.

One of the more fascinating points made throughout the book is the importance of building the right type of relationships. Nothing great has ever been accomplished alone! Our ability to relate to others in a personable manner, showing respect and treating employees with dignity will always provide a solid platform for building the right type of relationships.

This goes well beyond the basics because it takes a huge effort and most employers find this difficult to accomplish. It is even more difficult to maintain relationships with employees of larger companies; smaller companies can leverage their size because it is easier to get around to visiting their employees.

The greater the interaction with employees, the easier it becomes to build and nurture those relationships. It is important to realize that when employees trust their management team, they work harder, participate more, and get along with others better. That can lead to a safety culture that works well and experiences fewer accidents. Trust trumps all and employees will give you their best if they trust and believe you are working in their best interest.

Supervisors will need this guidance when they face the weight of responsibility that comes from managing workplace safety and risk. Management teams will be glad they took time to read this book; they will have a broader understanding of their roles and will be better equipped to support the organization's safety efforts.

This is a challenging time for small and mid-sized companies. Business owners, managers and community leaders are growing weary of the times; they need answers that will help them understand how to work with a new generation of employees. Todays' employees expect more, they are smarter, better educated and well informed due to the advances in education and technology!

We are in the midst of a rapidly growing global world and larger corporations usually have the resources to employ the professional staff they need to meet future challenges. On the other hand, smaller companies must hold it together and still meet those same regulatory requirements that are imposed by Federal, State and local governments.

One of the best ways to obtain buy-in for any program is to help employees visualize how their role supports the organization's vision. Having a clear vision and mission statement paints the canvas so that people can see where the company wants to go. People really do want to be a part of something larger than themselves and it is management's responsibility to connect the dots. A vision and mission statement is an important communication tool and should align the organization and its members.

The best organizations understand how important a system of communication is to their success. As an example, having a good safety management-training program communicates the importance of safety and is an investment in employees and in the organization's future. Investing in a formal safety-training program improves management's ability to counsel, teach and instruct others properly. Once managers are able to effectively communicate the organization's key objectives in terms that can be well defined, you have the means to fully engage your employees.

This takes work and you may need to ask some very tough questions. Why would anyone want to work for your organization? Do they enjoy the company, the people, the work, the pay, is it the economy or is there something else going on that you should be aware of that might provide more insight about the company?

Most employees look to identify with a product, quality or the company brand. The best companies are able to synchronize all three and find a way to build organizational pride. Pride of ownership is personal and people who are proud of their company tend to be the best marketers of the brand. They simply work harder and are more likely to work safely as well.

When it comes to a systematized approach to safety, health and the environment, the goals should be to solve a problem once. Then put an automated system in place that catches and utilizes previous experiences to prevent wasting time recreating solutions.

The final challenge is meeting regulatory requirements. This is something that can be managed internally or easily outsourced. There is an opportunity to lose money if you do not spend time selecting the best vendors or subcontractors who will meet your specific needs and provide you with quality services. Remember to keep your options open and don't worry over pricing so much that you get locked into long-term contracts that end up costing you more.

Of the eleven modern day risk factors that were identified in chapter three, the two most critically overlooked are safety philosophy and benchmarking. A simple change in philosophy and a focus on benchmarking would help most companies. Take a look at some of the steps that have been outlined here and perhaps you will uncover a new strategy or idea that you could use to make your job easier.

Remember that you lower your insurance premiums once you lower your incident rates!

There are plenty of opportunities to lower your rates if you manage your portion of the workers' compensation system well. Management teams must distance themselves from injury claims enough to allow claims investigators to remain objective.

Be careful not de-humanize an injured employee, push them aside or refer to them only as a case number as if they are unimportant. There is a very real risk of becoming too close to your work as well and that may prevent you from remaining objective. Moreover, once an accident investigation is underway, remember the process gets involved in peoples' lives and investigating accidents is tough work; use the system properly!

Management teams have access to information and technology that helps them arrive at decisions quicker; sometimes employees do not understand how managers can make such quick decisions about issues that will ultimately affect their lives. In some cases, this has caused employees to feel victimized. Managements' overriding goal should be to seek a balance between protecting the organization and the employee's right to fair and unbiased treatment.

It is imperative that your claims representatives do a great job of remaining connected to your employees. It is equally important that you stay in contact with the injured employee at the right intervals and maintain a caring and concerned relationship. This will help minimize any animosity that might otherwise spring up because this is a difficult time for the employee.

Most employees understand they are responsible for their behavior and that working safely in large part depends on them following your policies and procedures. Ultimately, safety is a matter of leadership excellence! The leadership expert and well-known author, _John C. Maxwell states that: "Everything Rises and falls on Leadership"._

Whether you decide to have an incentive plan, create a mastermind group or use consultants, it will eventually fall on how well you lead others! This book simply provides some insight from twenty-five years of practical experience. Success breeds confidence and getting great results will not only lower your risks, it will create the desire to continue winning! It will also reinforce and create the framework for the type of relationships you want to have with your employees.

When employees follow company policies and procedures without questioning the organization's motives and they trust the leadership, they become your safety champions. Once that occurs, employees will begin to believe the inevitable truth that: **_'Safety has no quitting time'._** Remember that safety must remain a 24/7 proposition and everyone must do their part for it to work. My new venture involves writing, speaking, training, teaching and coaching others. If you do what my lovely wife Susan once told me and "Take the leap"; you'll find that success is probably just on the other side of where you landed.

Whatever you decide to do, don't wait too long because the years are passing by ever so quickly. Take the leap, go out and offer the world the best of who you are and together maybe we can help others enjoy healthier, safer and more fulfilling lives. I wish you peace, happiness and your own personal success!

www.ingramcontent.com/pod-product-compliance
Lightning Source LLC
Chambersburg PA
CBHW071456210326
41597CB00018B/2574